FROZEN EARTH:

EXPLAINING
THE
ICE
AGES

R.V. FODOR

Assistant Professor of Geology
North Carolina State University

ENSLOW PUBLISHERS
Bloy Street and Ramsey Avenue
Box 777
Hillside, New Jersey 07205

Library of Congress Cataloging in Publication Data:

Fodor, R V
    Frozen earth.

      Adaptation of Ice ages, by J. Imbrie and K. P. Imbrie.
      "A Topical Press book."
      Bibliography: p.
      Includes index.
      1. Glacial epoch. 2. Glaciology. I. Imbrie, John. Ice ages.
II. Title.

QE697.F6   551.7'92   80-21588
ISBN 0-89490-036-6 (lib. bdg.)

Printed in the United States of America

10 9 8 7 6 5 4 3 2 1

CONTENTS

PREFACE

In 1971 the National Science Foundation organized the CLIMAP project to study changes in the climate of the earth over the past 700,000 years. One of the tasks was to make a map of how the earth looked during the last ice age. In the course of this assignment, the scientists examined cores of sediment from the Indian Ocean bed that gave a continuous geologic history of the earth for the past 500,000 years. Out of their analytical work came an exciting by-product: confirmation of one of the several theories about what causes ice ages.

Three of the principal scientists in the project are James D. Hays of Lamont-Doherty Geological Observatory, John Imbrie of Brown University, and Nicholas Shackleton of Cambridge University in England. It was their scientific paper, "Variations in the Earth's Orbit: Pacemaker of the Ice Ages," published in *Science*, that announced the results to the world in 1976. Subsequently, John Imbrie, together with his daughter Katherine, a science editor, published the book *Ice Ages: Solving the Mystery* (Enslow Publishers), presenting the discoveries in popular form. This book, which the prestigious journal *Nature* said "does for climate what J.D. Watson's *The Double Helix* did for molecular biology," won the 1979 Phi Beta Kappa Science Award as the best science book of the year.

Frozen Earth: Explaining the Ice Ages presents these recent findings about ice ages in a form accessible to students, and it is to them this book is dedicated.

The author wishes to thank John Imbrie for his encouragement during the writing of *Frozen Earth* and for reviewing the manuscript.

* * *

I have made acknowledgements in brief form in the captions to the illustrations. More extended versions of some of these sources appear in the bibliography. Sources that are too technical to appear in the bibliography are listed below.

R.V. Fodor

Emiliani, C., 1955, Pleistocene temperatures, *Journal of Geology, 63,* pp. 538-578.

Ericson, D.B., Ewing, M., Wollin, G., and Heezen, B.C., 1961, Atlantic deep-sea sediment cores, *Geological Society of America Bulletin, 72,* pp. 193-286.

Hays, J.D., Imbrie, J., and Shackleton, N.J., 1976, Variations in the earth's orbit: pacemaker of the ice ages, *Science, 194,* pp. 1121-1132.

Köppen, W., and Wegener, A., 1924, *Die klimate der geologischen vorzeit,* Gebruder Borntraeger, Berlin.

Mitchell, J.M., 1977, Carbon dioxide and future climate, *Environmental Data Service, March, U.S. Dept. of Commerce,* pp.3-9.

Shackleton, N.J., and Opdyke, N.D., 1973, Oxygen isotope and paleomagnetic stratigraphy of equatorial Pacific core V28-238: oxygen isotope temperatures and ice volumes on a 10^5 and 10^6 year scale, *Quaternary Research, 3,* pp.39-55.

1

BLANKETS OF ICE

It was midsummer. The skies were clear, and the sun's rays fell over most of North America. Nearly everywhere the wind was blowing. The day was icy cold even though it was summertime. The coldest winds blew off the mountains of ice that covered much of the land.

It was a good day to hunt. At the shore near New York, a polar bear and her cubs leaped from an ice raft and swam after seals. In Illinois, a pack of hungry arctic wolves trailed a musk ox along the edge of the ice. On the ice in Iowa, a furry arctic fox raced after a snow-white hare. And in Ohio, two men wearing animal skins walked the grasslands with heavy clubs, in search of reindeer.

Surely this is not a scene from last summer. Rather, it is a picture of a day 20,000 years ago—a time long before civilization, when cold-weather animals and early man roamed our land. Twenty thousand years ago was a time when a large part of the earth was covered by thick blankets of ice known as glaciers.

The surface of the earth looked greatly different then. Large parts of North America, Europe, and Asia were like today's ice-

Much of North America was under ice 20,000 years ago.

J.C. Holden

covered landmasses, Antarctica and Greenland. Ice up to a mile (1½ kilometers) high covered what is today the Great Lakes, the rich Midwestern farmlands, and the cities of Chicago, Cleveland, and New York. It covered all of New England. In the West, ice buried the mountains of Colorado, Wyoming, and California.

Glaciers have moved over the continents not just once but several times during the last million years. Each time they came, the seasons were cooler than they are today. On the average, days were colder by about 11 degrees Fahrenheit (6 degrees Centigrade). The cold weather and the advancing ice forced animals and early man to go to the southern parts of their lands.

Geologists, the scientists who study the earth, know that the most recent ice sheets reached their greatest sizes about

A long glacier streams down an Alpine valley in Switzerland.

Swiss National Tourist Office

20,000 years ago. Following that time, the climate warmed and the ice melted. By 7000 years ago, the glacier ice was gone, except for where it still is found today. That time 7000 years ago marked the end of a great ice age.

Discovering that there had been ice ages did not come easily for scientists because the ice sheets disappeared thousands of years before man began recording history. And the melting ice left few clues for the early geologists to find. Because of this, the study of glaciers and ice ages is a rather new science. It began in the early 1800s in Switzerland.

Switzerland is a fitting place for this science to begin because it is a land of mountains capped by snow and glaciers. More than 1200 glaciers dwell high in Switzerland's rocky Alps today. But it was actually in the green meadows and forests of

Switzerland's lower valleys that the first ideas about the ice ages developed. And rocks, not ice, first attracted the attention.

Geologists became curious about the many smooth and polished boulders that seemed to be lying out of place, scattered throughout the meadows and forests. Their presence was puzzling because these rocks were unlike those of the closest valley walls. The scientists of the time gave the name "erratics" to these out-of-place rocks.

Glaciers left these erratic boulders in Wisconsin during the last ice age.
Milwaukee Public Museum

One of the earliest explanations for erratics was that a great flood had carried them to the Swiss meadows. Because religion was an important part of the lives of many scientists, some believed that the erratic rocks were washed in from the Flood of Noah's time.

But that flood theory did not satisfy everyone. A famous

British geologist, Charles Lyell, argued strongly that *icebergs* once floating over Switzerland during a big flood had dropped the boulders. Sailors who had seen icebergs at sea reported that some do in fact contain boulders, so there was reason, then, to believe Lyell's icebergs-over-Switzerland theory.

Two scientists of the early nineteenth century who correctly said that the erratic boulders had been carried by glaciers were Ignace Venetz and Jean de Charpentier. Having visited and examined the glaciers, they saw that the boulders on the valley floors were similar to boulders in the high glaciers of the Alps. To them the explanation was simple: the erratic boulders had been moved into the valleys long ago when the glaciers had reached farther down from mountain tops. But could ice in the past really have flowed out of the high mountains, dropped boulders, and then retreated?

It took Louis Agassiz, a young Swiss naturalist, to convince scientists that ice actually could do that. Spending the summer of 1836 studying glaciers in the Alps, he learned enough about their movements to be certain that glaciers had carried the erratic rocks down from the high mountains to the lower valleys. Therefore, Agassiz believed, erratics in lower valleys meant that much more ice had once covered the earth than in the 1800s.

Louis Agassiz was an imaginative scientist. He became caught up in ideas about just how much glacier ice there had been in the past. After Agassiz decided he knew, he addressed a meeting of the Swiss Society of Natural Sciences. The time was July, 1837, and he told his audience that a gigantic ice sheet had once moved down from the North Pole and covered Europe. And without even having been to North America or Asia, he added that this great ice age had taken place there, too. Most of his fellow scientists did not believe a word of what he said.

Agassiz persisted. He took other scientists to the mountains

A rock surface in Canada that was grooved and polished by glaciers.

National Air Photo Library, Canada

and showed them where erratics had been left. He showed them where valley walls and floors had been scratched, grooved, and polished by ancient glaciers. And he showed them where huge mounds of dirt, known as moraines, had been bulldozed up and left by moving ice that later melted. Gradually Agassiz was able to prove to the leading scientists that he was right—that there had been an ice age.

In time, geologists found evidence of ancient glaciers in many parts of the world. They saw polished rock surfaces in New York State and in the Andes of South America. They even

These rolling hills in Ontario are moraines left by an ancient ice sheet.
Geological Survey of Canada

saw evidence for glaciation in the mountains of New Zealand and Australia. And geologists reasoned that if such great volumes of ice had developed, the water to make the ice must have come from the seas. Ocean levels must have been hundreds of feet lower during the Ice Age!

The scientists learned still other facts about what the earth was like during the Ice Age. Today's desert areas, for example, had been wetter and greener. Great Salt Lake in Utah today is only a small part of what was a huge lake during glacial times. And the earth's crust beneath thick ice sheets was pressed lower from the heavy weight of the ancient ice. Some areas, such as in Norway and Sweden, are still slowly springing back up to the levels they were at before the ice came.

Yet it was difficult for some scientists to believe that so much ice could have ever been in one place—enough to cover large parts of continents. An expedition to Greenland in 1852 helped remove those doubts, however. For the first time, scientists saw that an enormous sheet of ice can indeed occupy one place. By the late 1800s, then, there was barely a geologist around who did not believe that there had once been a great ice age.

One of the more important discoveries about ancient glaciers came in the 1850s. Geologic investigations in both England and in the Alps showed that the spread of glaciers was not a single event. Instead, there had been several advances and retreats of ice. Long cold periods with huge glaciers had alternated with long warm periods without many glaciers.

By the twentieth century, geologists knew that up to 4 ice ages had occurred during the past one million years. They also found that over one-quarter of all the land of the earth was covered by ice during the last ice age.

Recognizing those ice ages, however, created new problems to solve. The most mysterious one was how the climate of the earth could change so drastically as to bring on an ice age. Why does the world become so cold that mountain glaciers grow larger and spill into low valleys. And why do ice sheets enlarge and creep down from the polar regions? What causes an ice age—and then, why does it end?

But first, exactly what are glaciers?

2

GLACIERS - ICE ON THE MOVE

The winter season in many communities means living through months of cold weather and snow. Some places in North America, such as Minneapolis, Minnesota, and Winnipeg, Manitoba, Canada, have several weeks each year with temperatures below zero Fahrenheit (colder than 18 degrees below zero Centigrade) and with thick snow cover on the ground. In January and February, 1936, for example, Minneapolis had 36 straight days of temperatures dipping below zero. But no matter how frigid a winter is, both the cold and snow are always gone by summer.

There are some parts of the globe, however, where summer differs little from winter. The cold and snow remain all year. The high mountains of Alaska and Canada are examples of such places, as are the Sierra Nevadas in California. And landmasses near the poles of the earth—Greenland in the north and Antarctica in the south—have snow all year. In fact, more than a tenth of the earth's land has year-round snow.

Land where snow remains all year is said to be above the snowline. It is in these cold areas above the snowline that glaciers are born.

The snowline's height above sea level, or its elevation, is different from place to place across the earth. Snowline elevation depends greatly on the amount of heat a region receives from the sun (solar heat). For example, at the polar regions, where solar heat is relatively weak, the snowline is near sea level. It is even right at the sea in parts of Antarctica.

The elevation of the snowline also depends on how much snowfall an area receives. In the Northwest, snowfall is greater in the Olympic Mountains of Washington than in Glacier National Park of neighboring Montana. The snowline is therefore lower (closer to sea level) in Washington than in Montana.

Glaciers, however, are made of ice—not snow. But they do not form simply by water freezing as ice does on a pond.

Glaciers form above the snowline in the Olympic Mountains of Washington state.   U.S. National Park Service

Above the snowline, there is an additional process by which ice forms. It is called "compaction," which means pressing together.

Snowflakes buried under the heavy weight of hundreds of feet of overlying, younger snow are compacted so greatly that they become ice. This process resembles squeezing a handful of fluffy snow into an icy snowball.

Once a thick ice field develops, which may take hundreds of years, it begins to slowly move like thick molasses down any underlying slope. The moving ice is a glacier.

Every continent except Australia has glaciers today. There is enough ice in the world so that if it were spread out to cover all

A small valley glacier in the Olympic Mountains, Washington state.

National Science Foundation

An airplane view of the Antarctic ice sheet.

National Science Foundation

the land, the layer would be 400 feet (130 meters) deep.

The glaciers today are either valley glaciers (in mountains), or ice sheets spreading over the land as huge, cake-like masses. Valley glaciers, such as those in the Alps, move downward through mountain valleys as rivers of ice. Their movement is slow, however, often only a few inches a day, and rarely more than a few yards a day. The speed of a valley glacier depends largely on the steepness of the valley floor and on the thickness of the ice.

You can compare ice movement to that of a stream of

honey moving across a tilted plate. It you tilt the plate more or increase the amount of honey, the honey will flow faster. And the top portion will move faster than the bottom, which drags along the plate. In a glacier, the upper portion moves faster than the lower because the bottom drags against the valley floor and walls.

Measuring how fast the top of a glacier flows is easy. One common way is to pound stakes into the ice and then compare the positions of the stakes with markers on the valley walls each day.

One of the most unusual measurements of the speed of a glacier has to do with the tragic loss of a mountaineering group on a glacier in Switzerland in 1820. A snowslide buried an entire climbing party of Glacier des Bossens near the top of Mt. Blanc in Switzerland. Forty-one years later their bodies appeared at the foot of the glacier, about 10,000 feet (3000 meters) from the place of the accident. The glacier had carried the bodies about 8 inches (20 centimeters) a day.

Although glaciers move slowly most of the time, there have been times when some seemed to run out of control. One example was in the Alaska Range during 1937 when the normally sluggish Black Rapids glacier began to move over 100 feet (30 meters) a day. Several days it advanced nearly twice that much. The ice threatened to crush the Rapids Roadhouse (a hunting lodge) and to block the only road in and out of Fairbanks, Alaska.

The Revell family, living in the lodge at the time, reported that they heard the rumbling noises of the advancing ice every day and they felt it shake the ground. Then one day, several months after the rampage began, the glacier stopped, a half mile from the roadhouse. The pile of monstrous ice blocks at the front edge began to melt and disappear—but not before the glacier had stretched out of the mountains 6 miles (10 kilometers) farther than normal.

How far a glacier reaches before its front edge, or snout, melts depends greatly on the amount of snow being added to its source. The more snow that converts to ice high in the mountains, the longer the glacier is likely to become. In the case of the Black Rapids glacier, the sudden growth was due to several earlier winters of unusually heavy snowfall. Depending mainly on amounts of snowfall, then, glaciers today measure

Mountaineers preparing for a climb of Mountet glacier in Switzerland.
Swiss National Tourist Office

from less than 1 mile (1½ kilometers) in length, as in the Sierra Nevadas, to tens of miles long, such as the Hubbard glacier in Alaska.

Deep crevasses form in moving ice. *French Government Tourist Office*

Some valley glaciers travel long distances without melting away on land. They reach from mountaintops all the way down to the sea. Once entering the sea, ice breaks off, or calves, to float away as icebergs.

For as long as modern man has lived near glaciers, he has climbed and hiked across them. Mountain climbers and exploration parties frequently cross glaciers. In Switzerland, tourists visit glaciers simply to hike on them and to see what they are like. Usually, with spikes attached to boots and with ice ax in hand, a walk on a glacier can be done comfortably. But there is always one danger present: the crevasses, which are deep cracks in the ice.

As glaciers move down valleys, bending, and passing over humps on their valley floors, they break like gigantic pieces of plastic. Huge cracks develop that can reach hundreds of feet deep into the ice. Many mountain climbers have fallen to their deaths in these crevasses.

Ice sheets are the largest glaciers. The Greenland sheet, for example, is hundreds of miles long and up to 2 miles (3 kilometers) thick. Antarctica is covered by an ice sheet larger than the size of the United States and over 1 mile (1½ kilometers) thick in places. Three-quarters of the earth's fresh water is locked up in the Antarctic ice sheet. If it should ever melt, coastal cities, such as New York City, Los Angeles, and Miami, would be drowned.

Ice sheets move, too. Their directions of movement, however, resemble the spreading out in all directions from the center of an upside-down bowl, rather than movement in only one direction.

Most ice sheets spill into the sea. In one place where the Antarctic ice sheet meets the ocean, it floats outward as a huge shelf nearly 1000 feet (over 300 meters) thick and nearly the

U.S. Navy icebreakers remove an iceberg from the entrance to McMurdo Station, Antarctica. *U.S. Navy*

size of Texas.This ice mass is called the Ross Ice Shelf. Icebergs as big as mountains calve from the edge of this floating ice shelf. One berg measured larger than the state of Massachusetts.

Much of the geologic importance of glaciers lies in what happens at the base of the ice. The bottom of the ice picks up material of all sizes—from dust to boulders. This material then scrapes, scours, and polishes the rocks it passes over. The carving action is powerful. One study found that rock material in the Glacier des Bossens cut grooves 1½ feet (½ meter) long and 1½ inches (4 centimeters) deep in only one month.

Over a period of thousands of years, glaciers can create the greatest sculptures on earth. Deep U-shaped valleys such as those in Yosemite National Park of California and the fiords of Norway were cut by ancient glaciers. And jagged peaks, such as

Some of the best attractions at Yosemite National Park, California, are the valleys that glaciers carved during the ice ages.

U.S. National Park Service

the Matterhorn in Switzerland, were produced by glaciers that sculptured mountain tops.

The rocks used as scouring tools are transported and finally left in moraines piled at the melting end of the glacier. These moraines, along with erratic boulders and grooved rocks, provided early scientists like Louis Agassiz with proof that there had been ice ages. What the glaciers and glacial material have never revealed, however, is the cause of the ice ages. Yet since Agassiz first discussed ice ages in 1837, there has never been a shortage of ideas about what caused them.

3

ICE AGES: THE IDEAS WHY

Warm winds blowing across California from the Pacific Ocean are heavy with moisture. However, as they rise to cross the towering Sierra Nevadas, they cool and lose water as rainfall and snowfall. Then, by the time the winds pass eastward over the mountains to reach the Great Basin (eastern California, Nevada, and western Utah), they are depleted of moisture.

The change from moist to dry winds creates different climates in the western United States. That is, western California is wetter and less sunny than the Great Basin. And the high Sierras is cold and snowy. Having these different climates means that over a period of many years, each area has its own patterns of temperature, wind, rainfall, and sunshine.

Climate is important to ice ages. For ice to have grown and covered great portions of the earth's surface, the climate of the earth must have been much colder than it is today. Therefore, any explanation for ice ages must show why the climate of the world can be colder for long stretches of time.

A theory about ice ages must also take into account that once large ice sheets grew, this would have caused still more cooling of the climate. That is because the additional ice would have reflected more solar heat away fom the earth's surface, behaving like a mirror. With more heat lost, the climate would have become still cooler and the glaciers would have grown even more.

Such a chain of events, of course, would have been difficult to stop. Yet geologic history shows that something drastic did happen later to warm the climate, stop the growing ice sheets, and cause their retreat. The mystery to solve, then, is to find out what regularly changed the climate over the past one million years to turn ice ages on and off.

Because the sun's heat is the heart of our climate, nineteenth century scientists first looked to the sun for the answers. They reasoned that if the amount of solar heat, or radiation, reaching the earth had ever decreased in the past, the climate would have become colder. In turn, more snow would have

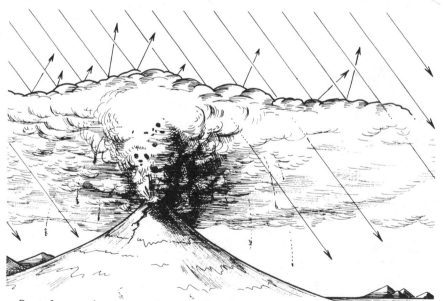

Dust from volcanic eruptions can be thick enough to block out sunlight.

J.C. Holden

fallen, the snowline would have moved lower, and glaciers steadily would have increased in size. But what could have suddenly decreased the amount of solar heat received on earth?

One possibility is volcanoes. People have known for a long time that volcanic eruptions can fill the sky with thick layers of cinders and dust. The eruption of Mt. Vesuvius in 79 A.D., for example, blocked out sunlight over Italy and turned daytime sky to blackness. And the volcanic eruption of Krakatoa in 1883 was so powerful that it destroyed half of the Indonesian island it occupied. People 3000 miles (5000 kilometers) away heard Krakatoa explode. So much rock dust was blasted into the atmosphere that sunsets all over the world were redder than normal for the following two years while the dust was gradually settling to the ground. Because the volcanic dust in the sky had blocked out some of the sun's radiation, temperatures across the world were cooler during those two years.

Could long periods of volcanic activity, then, have cooled the climate and triggered the growth of glaciers? Possibly, but modern-day scientists know of no unusually high amounts of volcanic activity in the earth's history. If there had been, many volcanic-dust layers would be mixed in the soils and in the muds on lake and ocean bottoms.

Perhaps changes in the sun itself caused a cooling of the earth's climate. This possibility has its roots with the Italian scientist Galileo, the inventor of the telescope, who lived in the early 1600s. He was the first person to use a telescope to report dark spots on the sun's surface. Although sunspots were mysterious in Galileo's time, they are now known to be cooler areas on the sun. However, their presence indicates greater overall solar activity. And that greater activity seems to cause changes in our climate.

Usually the greatest number of sunspots occurs every 11

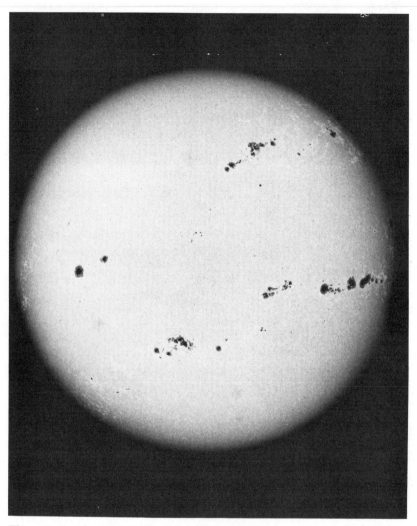

The activity of sun spots seems to affect our climate.

Hale Observatories

years. But there was one long period in history with almost no sunspots. In his research of sunspot records dating back to the 1600s, E. Walter Maunder of the Royal Greenwich Observatory in London at the turn of the century noted that sunspot activity was unusually low between 1645 and 1715. Maunder reported that the total sunspots for those 70 years was about as few as the number of spots usually seen in only one year's time.

At the time of that low sunspot activity, average temperatures were about 1 degree Centigrade cooler than they are now. That slightly cooler climate of the 1600s brought bitter cold winters to Europe and it changed the environment. Forests decreased in size, the snowline became lower, and glaciers in northern Europe grew and extended farther down their valleys than they do today. They became larger than they had been at any time since the last ice age, thousands of years earlier.

The creeping glaciers destroyed mountain huts and farmhouses and the highland meadows that livestock needed for summer grazing. The long-lasting colder seasons ruined crops and harvests. Problems with food, health, and shelter developed. These hardships forced thousands of people to leave the northlands of Europe.

That frigid period in history, now called the Little Ice Age, lasted until the mid-1800s. Today, some scientists use the facts about the Little Ice Age to develop an explanation for the great ice ages. They say that a change in solar activity, as indicated by a low number of sunspots over a long period of time, is the chief cause of former colder climates and glaciations. That theory, however, is difficult to test or prove. There are no records of sunspots for the times there were ice ages.

The movement of our solar system (the sun together with its planets) is the basis for another theory about how the sun's behavior could trigger ice ages. The theory is that at times, the sun enters enormous clouds of dust in space. This dust, falling into the sun, disturbs its normal activity. In turn, the earth's climate is affected. The changes, some believe, could lead to

The Rhone glacier in Switzerland was much longer during the Little Ice Age of the 1800s (above) than it is today (opposite).

Swiss National Tourist Office

the formation of ice clouds in our atmosphere that would prevent some of the sun's heat from reaching the earth's surface.

But this idea, too, of the sun passing through interstellar dust, cannot be tested or proven as the cause of ice ages. Part of the difficulty is that scientists do not actually know what exactly happened to our atmosphere and climate when (and if) the sun traveled through dust clouds.

Another possible cause of ice ages lies closer to the earth, in the atmosphere. The condition of the earth's atmosphere greatly affects climate. Should it become richer in carbon dioxide, for instance, the climate would warm up. This would happen because carbon dioxide (plus water vapor and clouds) traps heat radiating off the earth's surface. The more carbon

dioxide there is, the more solar energy the atmosphere retains. Carbon dioxide, then, can make the atmosphere behave as a one-sided heat shield, much as the windows do on a greenhouse. Heat becomes trapped inside. Too much carbon dioxide in the atmosphere could cause the earth to reach what would be uncomfortably high temperatures for man. Venus has this "greenhouse effect." That planet is surrounded by a carbon dioxide atmosphere that helps raise surface temperatures to about 500 degrees Centigrade.

Unfortunately, the greenhouse effect may someday become a problem on earth. By burning fuels such as gasoline,

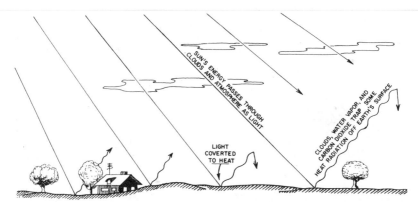

Carbon dioxide in the atmosphere helps to trap heat near the earth's surface.   J.C. Holden

oil, and wood, we are continually adding carbon dioxide to the atmosphere.

   If there were times in the past when there was little carbon dioxide in the atmosphere, the reverse would have happened. Too much heat would have escaped through the atmosphere and the climate would have cooled to bring on an ice age. Missing from this ice age theory, however, is a reason why carbon dioxide could suddenly have become unusually low.

   No explanation for ice ages has received more attention than one based on how the earth travels around the sun and spins on its axis. Although this theory originated during Agassiz's time, it actually extends back to 120 B.C. That was the year that the Greek astronomer Hipparchus discovered that the earth's North Pole has not always pointed the same direction in space.

   During recent centuries the North Pole has pointed to the North Star, the star at the end of the handle of the Little Dipper constellation. The ancient navigators used that knowledge to find their directions at sea. But 4000 years ago,

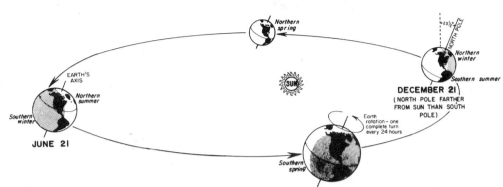

The tilt of the earth's axis is the reason for the seasons. The Northern Hemisphere has winter when the North Pole tilts away from the sun.

J.C. Holden

the North Pole pointed halfway between the Little Dipper and the Big Dipper. And 6000 years ago, the North Pole pointed toward the top of the Big Dipper's handle. Throughout time, then, the top of the earth (the North Pole) has been "wobbling," pointing in different directions.

Both the North and South poles are the ends of an imaginary axis about which the earth spins. One turn around this axis is equal to about 24 hours, a day and a night. The axis, however, is not straight up and down. Rather, it is tilted. Mathematically, the tilt is at an angle of about 23½ degrees—commonly written as 23½°. It is this axis that points today from the North Pole to the North Star.

The 23½° tilt of the earth is the main reason for the seasons. When the North Pole tips away from the sun, the season is winter. At the same time, the South Pole is closer, tipping toward the sun and bringing summer to the Southern Hemisphere. Similarly, when the North Pole tips toward the sun, the Northern Hemisphere has summer.

What Hipparchus determined 2000 years ago was that the earth's axis "wobbles" like that of a spinning top slowing down: at one time the top can tilt away from you; at another time,

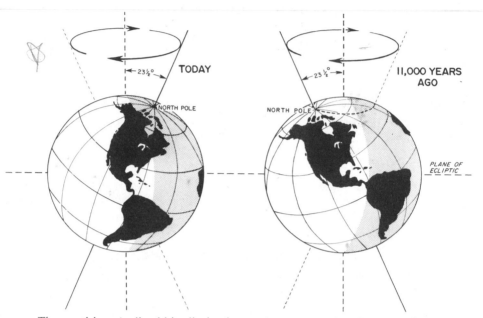

The earth's axis "wobbles," slowly carving out an imaginary circle in space once every 22,000 years.   J.C. Holden

it can tip toward you. The top of the earth's axis, then, moves about "drawing" an imaginary circle in space. This movement is called axial precession. The time it takes to complete the circle is very long—about 22,000 years.

What Hipparchus did not know, however, was that this slow wobble affects climate. That discovery came much later. And it set the background for the only theory proven to be a cause of ice ages: the astronomical theory.

4

THE ASTRONOMICAL THEORY

In the seventeenth century, Johannes Kepler made a striking discovery about how the earth travels around the sun. The earth's orbit, he calculated, is not a perfect circle. Rather, the path of the earth's yearly trip around the sun is an ellipse— a somewhat out-of-round, or stretched-out, circle.

Because of this elliptical orbit, the earth is closer to the sun at a certain time each year than at any other time. It is also farther from the sun at a certain time each year than at other times. Presently, the orbit takes the earth closest to the sun during winter (for the Northern Hemisphere) and farthest from the sun in summer.

Two centuries after Kepler, the French astronomer Urbain Leverrier discovered that the shape of the orbit gradually changes over thousands of years. It goes from a pronounced ellipse, to nearly a circle, and back to a pronounced ellipse again. This happens because of the gravitational pull on earth from other planetary bodies in the solar system. The amount of elongation in the orbit (how much it is stretched) at any time is called the orbital eccentricity.

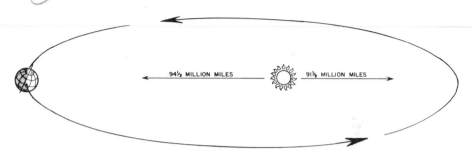

94½ MILLION MILES   91½ MILLION MILES

The earth's orbit is not a perfect circle, but rather, an ellipse.

J.C. Holden

Today the earth's orbit is closer to being a circle than a true ellipse. It has low eccentricity. But thousands of years ago the orbital eccentricity was high, as it will be again thousands of years from now.

In the 1860s, the Scotsman James Croll showed that the change in the earth's orbit from high to low eccentricity, and back to high again, takes about 100,000 years. That is, approximately every 100,000 years, the orbit is in a period where it carries the earth unusually far from the sun for part of each year.

What if, Croll wondered, the axial precession (the "wobble" of the earth's axis) put the earth's axis tilting directly away from the sun at the same time that the orbit was highly eccentric and carrying the earth unusually far from the sun once

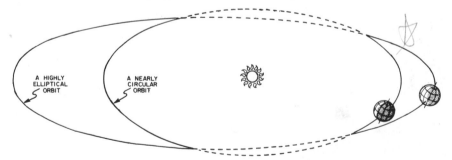

A HIGHLY ELLIPTICAL ORBIT   A NEARLY CIRCULAR ORBIT

Over long periods of time, the shape of the earth's orbit changes between highly elliptical and nearly circular.

J.C. Holden

a year? This should create the coldest winters possible. And if winters are very cold, more snow will accumulate. That, in turn, will reflect more of the sun's heat, to gradually cool the climate. These conditions should create an ice age.

Croll used astronomy, then, to explain ice ages. When the earth's orbit was in a period of high eccentricity, either the Northern Hemisphere or the Southern Hemisphere would have an ice age. Whether it was one or the other would depend on which hemisphere had its winter (the season caused by the pole tilting away from the sun) while at the farthest place in the orbit from the sun.

Croll combined the 22,000-year axial precession cycle with the 100,000-year orbital cycle (orbital eccentricity) to account for cooler climates. He believed that once these astronomical

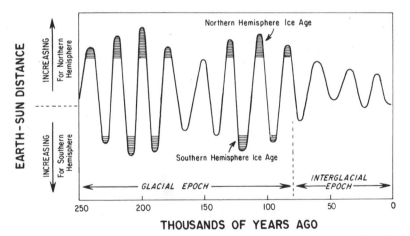

Croll's charts showed that many ice ages have occurred in the Northern and Southern hemispheres, but at alternating times. Important in his idea was a pole (or hemisphere) being near its greatest possible distance from the sun.

Adapted from Imbrie and Imbrie, 1979

events cooled the climate, an ice age would follow and last about 10,000 years. And through his mathematical calculations, he showed that over the period from 80,000 to 250,000 years ago, the Northern and Southern hemispheres had many alternating ice ages and warm periods. Croll called an ice age period a "glacial epoch." And he called the warmer times between ice ages "interglacial epochs."

For a while, Croll's explanation for ice ages appealed to his fellow scientists. But it eventually met defeat because some glacial tills (layers of soil and rock left, or deposited by ice-age glaciers) were found to be much younger than 80,000 years old. In fact, geologists estimated some tills to be only 10,000 years old. Because Croll's work did not show evidence for such a recent ice age, his astronomical theory was laid to rest.

About twenty years after Croll's death in 1890, the Yugoslavian mathematician Milutin Milankovitch revived the astronomical theory for explaining ice ages. Like Croll, Milankovitch also was interested in ancient climates and in the orbital path of the earth. His chief interest was in how much solar heat, or radiation, had reached the regions of the earth between the equator and the North Pole in the past. He worked several years on mathematical calculations to obtain this information. When he finally published his work in 1920, he was able to demonstrate that astronomical events could have caused ice ages by influencing how much solar radiation fell on the earth in ancient seasons.

Milankovitch's work stirred up new interest in the astronomical theory as the cause of ice ages. Where he mainly differed from Croll before him was in recognizing the role of axial tilt. Croll knew of the 22,000-year period for axial precession, but he did not consider that the earth's tilted axis is not fixed at 23½°. That is, over time, the tilt slowly increases

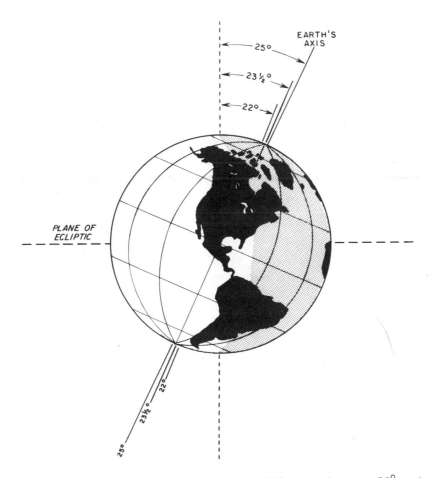

The amount of axial tilt changes over 41,000 years between 22° and 25°. Today it is at 23½°. J.C. Holden

to about 25°, and then decreases to about 22°. Astronomers call this 3° shift in tilt the change in the obliquity of the ecliptic. The complete cycle for the shift takes about 41,000 years.

 There were now three astronomical events to include in the explanation of ice ages: (1) eccentricity of the orbit, with a cycle of about 100,000 years; (2) axial precession, with a cycle of about 22,000 years; and (3) the change in the obliquity of the ecliptic, with a cycle of about 41,000 years.

But Milankovitch believed that only axial precession and axial tilt were important for drastically changing climates. By 1930, he had calculations and graphs and charts to show how these two events had caused summer radiation at 8 zones, or latitudes, on the globe to vary over the past 650,000 years. For example, the high radiation period between 300,000 and 400,000 years ago marked a time of little ice and high sea level. And the low radiation period between 175,000 and 250,000 years ago had ice ages and low sea level.

Milankovitch had no proof for his ideas about climatic changes, however. They were based only on calculations. Finding proof was up to the geologists.

Some scientists noted that the 4 ice ages that had been observed from glacial tills in North America equalled the 4

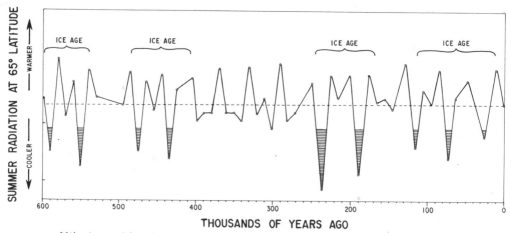

Milankovitch's charts showed four ice ages over the past 600,000 years. Important here is how much summer radiation the earth received, as controlled by axial precession and axial tilt. The low radiation points on this chart for 65° latitude represent cool climates.   Adapted from W. Köppen and A. Wegener, 1924 times of low summer radiation that Milankovitch showed. They accepted this as proof, even though there were no accurate age-dates for the tills. That is, no one knew exactly how many years ago the tills were deposited by ice.

Many geologists in Europe were also quick to support Milankovitch. In fact, some became carried away and began using Milankovitch's charts to age-date the glacial tills. To them, the ice-age mystery was solved.

But it wasn't. Other scientists were cautious and unwilling to settle for the astronomical theory—no matter how good it seemed. The greatest problem was that there was no way to reliably compare the ages of glacial material such as tills, to the times Milankovitch showed by his charts to be cooler.

Eventually, then, scientists set Milankovitch's astronomical theory aside. It awaited proof or disproof from geologic material that may have recorded what the climates were like at the times that the theory predicted there had been ice ages.

The real downfall for Milankovitch came in the 1950s when scientists began using the radioactivity of the element carbon to determine ages of glacial material. Carbon age-dating showed that his astronomical calculations were probably of little value. Some age-dated material indicated that there had been ice ages during periods Milankovitch thought to be warm. Other age-dated material suggested warm climates for when Milankovitch had predicted ice ages.

Again the astronomical theory—long the most promising explanation for ice ages—met defeat. By 1965, the theory was practically finished. And once again, the reasons for ice ages became a mystery.

But as it turned out, Milankovitch's calculations had not received fair treatment. The arguments against them were based on an incomplete climatic record taken from the land surface. Had geologists of the time been able to investigate the seafloor records for ancient climates, they might have learned that Milankovitch and his calculations were winners after all.

5

SOLVING THE

ICE-AGES MYSTERY

Two weeks before Christmas in 1872, the 223-foot (70 meter) warship *H.M.S. Challenger* left from England on a four-year journey around the world. Combat was not its mission, however, for the guns had been removed and scientific equipment and laboratories added. Instead, the *Challenger's* crew planned to explore ocean depths. It hoped to find answers to questions about the sea that people had asked for thousands of years.

That scientific voyage marked the birth of oceanography. The *Challenger* traveled 69,000 miles (111,000 kilometers) measuring water depths and collecting plants, animals, rocks, and sediments (clay and mud on the seafloor) at 365 stations. The trip provided the scientific community with enough material and information to occupy twenty years of research time. For example, biologists discovered many new types of sea life, and geologists learned of huge mountain ranges crossing ocean bottoms.

Climatologists also profited from the *Challenger's* haul. The excursion produced the first clues that oceans contained

H.M.S. Challenger.   *Reports of the Challenger*

information about ancient climates. The discovery lay in tiny floating animals called forams—and their fossilized remains in seafloor sediments.

But to reconstruct the climatic history, scientists had to recover thick accumulations of foram layers from beneath sea water miles deep. What they needed were cores of the seafloor. A core of seafloor material would show the different layers of sediments and fossils; the youngest would be at the top of the core and the older at the bottom.

The tool first used to gather seafloor sediment was a heavy pipe. A ship crew lowered it into the water until it hung just above the seafloor. When released, the pipe plunged into the sediment, forcing it up into the core of the pipe. The pipe and its sediment filling then were raised aboard ship.

Geologists examining halves of sediment core recovered from the Atlantic Ocean floor. *Woods Hole Oceanographic Institute*

This coring worked, but the scientific information that the cores provided was limited. In many cases, the sediment cores were too short, 1 meter or less. They therefore did not produce enough information to help unravel climatic history and to test Milankovitch's theory for ice ages. In other cases, the sediment layers were disturbed and mixed up too much by the plunging of the pipe.

Yet by 1935, studies of many cores from the Atlantic Ocean showed a pattern. The top sediment layer contained fossil forams different from those in the underlying layer. The top, Layer 1, contained "warm" water forams like those presently living in the sea. Layer 2 beneath was composed mainly of what were believed "cold" water forams. Farther down, "warm" water forams occurred again as part of Layer 3.

One foram species in particular seemed to be an outstanding indicator of warm-water temperatures. Scientists call it *Globorotalia menardii*. The absence of *menardii* in Layer 2 indicated that the sedimentary layer accumulated when the waters were cooler than they are today—during an ice-age climate.

But scientists needed still longer cores, perhaps 10 to 15 meters long, for looking farther back in time. Also, if they were to prove or disprove Milankovitch's charts for interglacial and glacial periods, they needed a way to age-date the different "warm" and "cold" sediment layers they recovered from the sea.

The recovery of long cores finally was achieved in the late 1940s. The new technique used a tube that sucked up sediments while it was being forced into the ocean floor. Research ships at sea then went full-swing into a program of collecting as many long cores as possible. Some ships collected up to 200 cores a year. Most of the cores were stored at Lamont-Doherty Geological Observatory (of Columbia University) in Palisades, New York, and at Scripps Institution of Oceanography (University of California at San Diego) in La Jolla, California.

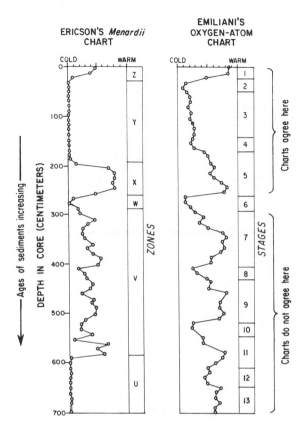

ERICSON'S *Menardii* CHART

EMILIANI'S OXYGEN-ATOM CHART

The charts of Ericson and Emiliani do not agree well as to when the different sediment layers represent warm and cold periods.

Information from C. Emiliani, 1955, and Ericson et al., 1961

As expected, the longer cores contained more layers, both with and without *menardii*, the warm-water fossil. David B. Ericson, working at Lamont-Doherty Geological Observatory, was confident that the alternating layers defined alternating glacial and interglacial periods. According to Ericson, the ice ages left their marks in the seafloor in the form of layers with few *menardii*.

Ericson's work seemed to bring geologists one step closer to understanding the ancient climatic record. But then suddenly, a new discovery raised more questions.

In 1955, Cesare Emiliani, then at the University of Chicago, developed a method to determine temperatures of ancient oceans. It required examining the types of oxygen atoms present in seafloor fossils. When applied to the cores

A microscopic view of Globorotalia menardii, *a foram. It is one milli-meter in size.*   *A. Bé*

examined by Ericson, Emiliani's oxygen-atom charts, like Ericson's charts, showed that there had been several glacial-interglacial periods over the past 300,000 years.

But to Emiliani's disappointment, the details of his work strongly disagreed with Ericson's research. The oxygen "thermometer" showed some warm zones where Ericson's work demonstrated cold zones. If scientists were to test

Milankovitch's astronomical theory, they had to agree on which sediment layers showed when past periods were warm and when they were cold.

Scientists began to question the research methods. For example, Ericson wondered if Emiliani's oxygen thermometer was accurate. Others wondered if foram populations were really that sensitive to temperature changes. The uncertainties were so great that the National Science Foundation held a special conference in 1965 just for the purpose of resolving the conflict between Emiliani and Ericson. But no resolution was reached. The road to solving the ice-age mystery by using seafloor material seemed to reach a dead end.

The barrier turned out to be temporary, however. The following years brought scientific developments that enabled the search for proof of the astronomical theory to continue. They also brought explanations for why Ericson's and Emiliani's research results disagreed.

Throughout the 1960s, John Imbrie of Brown University and Nicholas Shackleton of Cambridge University had developed new theories about forams. Their work showed that whether or not *menardii* were present as fossils depended on more than just cold or warm water. Salt content of sea water also affected foram populations. Imbrie and Shackleton also found that Emiliani's oxygen-atom research told more about the sizes of global ice sheets than it did about water temperatures.

These were major discoveries. Knowing whether ancient sea water was saltier than today (saltier water means that more fresh water is tied up in glaciers), and therefore whether the amount of ice on the globe is large or small, is actually better than having a "thermometer" for ancient sea water.

At the same time, James D. Hays and Neil D. Opdyke of

Lamont-Doherty Geological Observatory worked on improving the geologic time scale. They combined a new understanding of the magnetic properties of seafloor material with knowledge about the magnetic field of the earth. By 1971, their results enabled accurate age-dating of sediment-core layers. The age-dating could tell when (how many years ago) the different layers accumulated on the seafloor. This information, combined with knowing which layers were "warm" and which were "cold," could test the astronomical theory.

Elsewhere, scientists examined ancient beaches called terraces. These are coastlines that have been sculptured by waves and now look like giant flights of stairs. Terraces show that sea levels have been much higher at times in the past, such as around 80,000 years, 105,000 years, and 125,000 years ago.

Terraces along a Pacific Ocean coast, New Guinea. Each level represents a time of high sea level in the past.   A. Bloom

These times are close to the times Milankovitch believed to be interglacial, or when most of the earth's water would be in the sea rather than frozen on land as ice. Learning this gave a strong boost to Milankovitch's astronomical theory.

Also in the 1960s, new information came from studying certain soils that had formed during the ice ages. In Czechoslovakia, for example, soils showed that ice ages came on slowly but ended quickly. And both new deep-sea cores and certain soils showed that ancient climates had shifted according to 100,000-year cycles.

The discovery of these 100,000-year cycles brought back memories of Croll's astronomical theory. You will recall that this Scotsman believed orbital eccentricity to play a major role in climatic changes. Milankovitch, on the other hand, had thought that axial tilt and axial precession were the key parts of the astronomical theory for climatic changes.

Still, with all this new information learned in the 1960s, scientists could not make any firm conclusions. There were no final answers even though an accurate age-dating method was available for sediments, and even though oxygen-atom analyses of sediments could distinguish when the volumes of ancient ice were large or small.

One puzzle piece remained missing: scientists needed to apply their age-dating and oxygen-atom work to a special core. It had to be composed of sediments that had accumulated rapidly enough to reveal very detailed changes in climate. And it had to be long enough to contain sediments hundreds of thousands of years old.

Part of this need was finally met in a core raised from the Pacific Ocean in 1971. Age-dating and oxygen-atom examinations of this core showed that the climate had changed from cold to warm and back again 19 times over the last

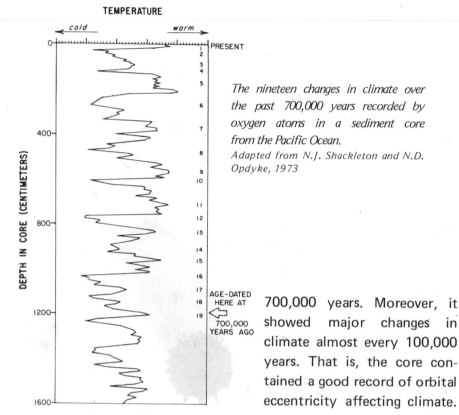

TEMPERATURE

The nineteen changes in climate over the past 700,000 years recorded by oxygen atoms in a sediment core from the Pacific Ocean.
Adapted from N.J. Shackleton and N.D. Opdyke, 1973

700,000 years. Moreover, it showed major changes in climate almost every 100,000 years. That is, the core contained a good record of orbital eccentricity affecting climate.

Some of the evidence was now in. But scientists still needed proof that axial tilt and axial precession affected climate. After hundreds of core examinations, however, the chances of finding their cycles recorded in a sediment core seemed smaller than finding a needle in a haystack.

Then, in 1973, James D. Hays discovered the needle. It was a core from the southern Indian Ocean that was stored at Florida State University. Its sediment layers dated back 450,000 years. Also, the layers had accumulated on the seafloor rapidly enough to record detailed climatic changes. Hays found not only the 100,000-year climatic cycle in the core, but also evidence for climatic changes at the times Milankovitch had predicted from information about axial tilt

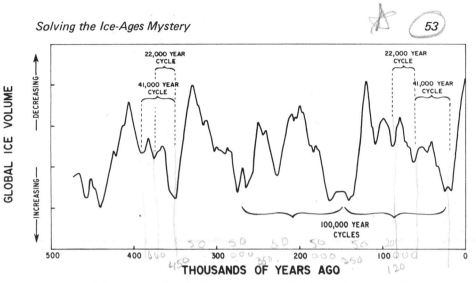

One core taken from the Indian Ocean in 1971 records several changes in ice volume, many of which agree with the 100,000-, 41,000-, and 22,000-year cycles of orbital eccentricity, axial tilt, and axial precession. The cycles are not always perfectly spaced because the three events (eccentricity, tilt, and precession) sometimes work to offset each other. Note that eccentricity has the greatest effect on cooling the climate.

Adapted from J.D. Hays et al., 1976

and axial precession. These were the 41,000-year and 22,000-year cycles.

All evidence was finally in. By the mid-1970s, scientists could state with confidence that ice ages came and went over the past million years mainly because of changes in climate due to orbital eccentricity, axial tilt, and axial precession. Seafloor sediments record that each of these astronomical cycles works to change the earth's climate. The cycle that seems to have the strongest effect is the orbital eccentricity one.

As so often happens in science, solving a problem raises new questions. In the case of the ice-age mystery, finding evidence for the astronomical theory brought questions about what to believe about the other ideas scientists had for the ice ages. For example, should the volcanic-dust and interstellar-dust theories now be ruled out because of the evidence for the astronomical theory? What about sunspots? Can they affect climate?

The whole ice-age problem is indeed complex. Even though astronomical changes are a main cause of climatic changes and therefore, of ice ages, other events have certainly affected climates, too. That is, changes in sunspot activity and the amounts of volcanic dust and carbon dioxide in the atmosphere also may have contributed to climatic changes in the past, even if only in a small way. But the astronomical theory differs from these other explanations by being the only one for which there is *evidence* as a cause of ice ages.

Two other questions still have to be answered. Since astronomical events take place over and over again, have ice ages occurred regularly throughout all of the history of the earth? And will there be ice ages in the future?

6

ICE AGES: EARLIER STILL

–AND STILL TO COME

During the ice ages, the largest glaciers formed in the earth's high latitudes. These are the regions near the North and South poles. Even today, huge ice caps occupy the high latitudes. They cover Greenland in the high north, and Antarctica in the high south.

But suppose that in the geologic past there was no land near the poles on which glaciers could form. Would there have been ice ages? And suppose that *today* there was no land near the poles. Would there be ice sheets today?

The answers are no. Ice sheets would not form without land near the poles. That is probably why geologic history does not include a record of regularly alternating ice ages and interglacial ages. There has not always been land near the poles!

Over the past two decades, geologists have gathered evidence that the continents and ocean floors are not stationary. Rather, they are parts of several huge "plates" about 60 miles (100 kilometers) thick. These plates move past, over, and under one another across the globe. Their movement, however, is slow—about 1 inch (2½ centimeters) a year. Yet

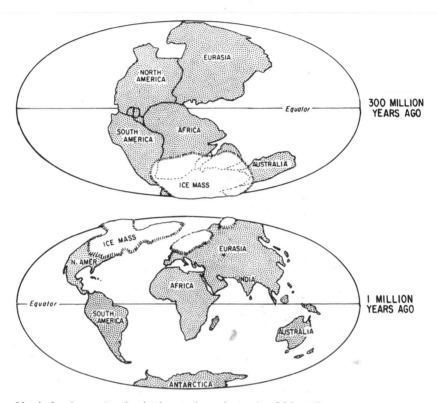

Much land was in the high southern latitudes 300 million years ago and ice ages occurred in the Southern Hemisphere. With much land in the high northern latitudes one million years ago, ice ages occurred there.

J.C. Holden

this is enough to cause earthquakes, the eruption of volcanoes, and the building of mountains.

Scientists now believe that about 300 million years ago, all the continents we know today were grouped together as one supercontinent, called "Pangaea" (a Greek word meaning "all earth"). Much of Pangaea was located in the Southern Hemisphere, whereas little of it was in the northern high latitudes.

About 200 million years ago, however, Pangaea began to

break apart. Over the time since, the landmasses have slowly drifted to their present positions. The Atlantic Ocean filled in where North and South America broke from Europe and Africa. Currently, there is ample land in the high latitudes of both the Northern and Southern hemispheres on which glaciers can form.

If landmasses must be in certain geographic locations for ice ages to occur, the earth's geologic record for the past 300 million years should contain few periods of alternating ice ages and warm intervals (periods of glaciation). As it turns out, the geologic record shows ice ages for the period 250 to 325 million years ago, when South America, South Africa, India, and Australia were at high southern latitudes, and for the last few million years, when North America and Eurasia have occupied the northern latitudes.

Even though climate records show ice ages for about only the past one million years, this period of glaciation actually began millions of years earlier. Geologists see evidence of glaciers having begun in the high latitudes (such as in Alaska) 5 to 10 milllon years ago. And 3 million years ago, ice sheets appeared in the Northern Hemisphere. Once formed, their growing and shrinking seem to have been sensitive to the astronomical variations.

A deeper look into geologic history shows one other great period of glaciation that occurred 600 to 800 million years ago. The positions of continents that far back in time are not clear, but most likely the glaciated portions were in high latitudes.

The movements of the continents, then, can explain why glaciations began in the first place. If landmasses were in high latitudes when orbital eccentricity, axial tilt, and axial precession acted to create cool climates, ice sheets would have developed. The astronomical theory, on the other hand, answers the question of why ice ages turned off and on once a glaciation period began.

But what of the future: are we due for another ice age soon?

The geographic locations of certain plants and animals have changed over the past several thousand years. For example, certain species of oak trees and mussels once abundant in Scandanavia now are absent there. Elsewhere in Europe, certain vegetation zones have moved southward. These changes in habitat point to a climate that has been gradually cooling over the past few thousand years.

The cooling is not steady, however. Rather, there are short episodes of intense cooling, like the Little Ice Age of 1650 to 1850, mixed with short warming periods. But the overall change is toward a cooler climate.

The astronomical theory can predict when world temperatures will reach their lowest. Presently, eccentricity and tilt are working to cool the climate; axial precession is warming it. In combination, these events will lead to a very cool climate about 10,000 to 15,000 years from now. Ice sheets will once again grow large, and the earth will be in the depths of another ice age about 20,000 to 25,000 years from now.

Before that ice age begins, however, there will be a super-interglacial age. It will be an extension of our present interglacial time and marked by temperatures climbing like they never have before. Curiously, the cause of this will be man.

Superwarmth will come from the burning of coal and oil over the next 100 years. Industries and private citizens already pump more than 20 billion tons a year of carbon dioxide into the atmosphere. By the next century, the amount may be 40 billion tons a year. This will produce a strong greenhouse effect that would last for the following 1000 years.

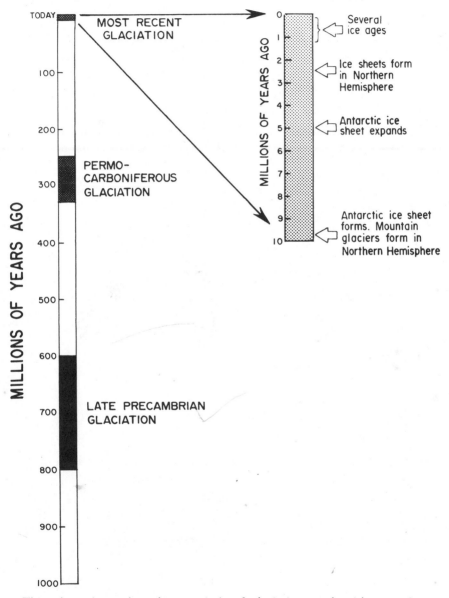

There have been three long periods of glaciation, each with many ice ages, over the past one billion years.   J.C. Holden

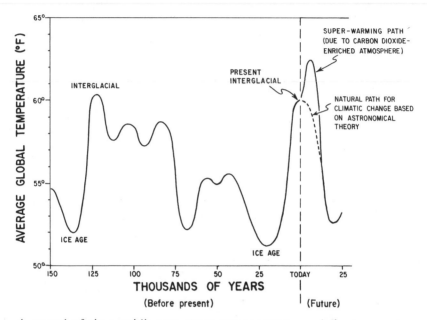

A record of the world's past average temperature, and the temperature changes for the future. By man adding carbon dioxide to the atmosphere, the climate will not follow its natural course according to the astronomical theory.

Modified from Mitchell, 1977

During this time, temperatures will rise 4 to 5 degrees Fahrenheit (2° to 3° Centigrade).   Much of our present ice sheets will melt and raise the sea level. Coastline cities and farmlands will be drowned. The added moisture will change the climate enough to bring some deserts to bloom. But when the superwarmth ends, the climate will continue its slide into still another ice age, some thousands of years from now.

BIBLIOGRAPHY

Brooks, C.E.P., 1970, *Climate through the ages,* Dover, New York.

Cornwall, Ian, 1970, *Ice ages: their nature and effect,* John Baker, London.

Chapman, R.D., 1978, *Discovering astronomy,* W.H. Freeman, San Francisco.

Flint, R.F., 1971, *Glacial and quaternary geology,* John Wiley, New York.

Fodor, R.V., 1978, *Earth in motion: the concept of plate tectonics,* Morrow, New York.

Frenzel, B., 1973, *Climatic fluctuations of the ice ages,* Press of Case Western Reserve University, Cleveland.

Imbrie, J., and Imbrie, K.P., 1979, *Ice ages: solving the mystery,* Enslow Publishers, Hillside, NJ.

John, D.S., 1977, *The ice age: past and present,* Collins, Glasgow.

Lewis, R.S., 1979, *The coming of the ice age,* Putnam, New York.

Lurie, E., 1960, *Louis Agassiz: a life in science,* University of Chicago Press, Chicago.

Matsch, C.L., 1976, *North America and the great ice age,* McGraw-Hill, New York.

Nixon, H.H., and Nixon, J.L., 1980, *Glaciers: nature's frozen rivers,* Dodd, New York.

Schultz, Gwen, 1974, *Ice age lost,* Anchor/Doubleday, Garden City, New York.

Sharp, R.P., 1960, *Glaciers,* Oregon State System of Higher Education.

Wyatt, S.P., 1977, *Principles of astronomy,* Allyn and Bacon, Boston.

INDEX